# Maths
# Frameworking

Chris Pearce

Intervention Workbook

William Collins' dream of knowledge for all began with the publication of his first book in 1819. A self-educated mill worker, he not only enriched millions of lives, but also founded a flourishing publishing house. Today, staying true to this spirit, Collins books are packed with inspiration, innovation and practical expertise. They place you at the centre of a world of possibility and give you exactly what you need to explore it.

Collins. Freedom to teach.

**Published by Collins**
An imprint of HarperCollins*Publishers*
77–85 Fulham Palace Road
Hammersmith
London
W6 8JB

Browse the complete Collins catalogue at
www.collins.co.uk

**Acknowledgements**
The publishers wish to thank the following for permission to reproduce photographs. Every effort has been made to trace copyright holders and to obtain their permission for the use of copyright materials. The publishers will gladly receive any information enabling them to rectify any error or omission at the first opportunity.

Cover Carl Schneider/Getty Images

British Library Cataloguing in Publication Data
A Catalogue record for this publication is available from the British Library.

Commissioned by Katie Sergeant
Project managed by Elektra Media Ltd
Developed and copy-edited by Joan Miller
Proofread by Amanda Dickson and Joan Miller
Edited by Helen Marsden
Illustrations by Ann Paganuzzi
Typeset by Jouve India Private Limited
Cover design by Angela English

With special thanks to Maxine Meyer and Ian Nicholls.

Printed and bound by L.E.G.O. S.p.A, Italy

# How to use this book

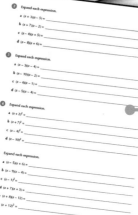

## Organise your learning

The Contents table at the start of the Workbook shows the topics you are going to cover. It can be used by:

- your teacher or tutor can set a date for you to complete each topic by
- you can give a traffic light colour for each topic to show how you feel it went
- you, your teacher and your parent or carer to write comments.

## Work through each topic step by step

For each topic, there are:

- clear learning objectives
- worked examples to show you how to answer the questions
- practice questions to help you consolidate what you have learnt. A glossary and answers are available on the Collins website.

At the end of each chapter, there's a comments box for your teacher or tutor to fill in on how you did.

## Practise your mental maths

Try the mental maths questions at the end of the Workbook to see what you have learned.

## Celebrate your progress

When you finish the Workbook, your teacher or tutor can fill in the Record of achievement certificate for you to keep.

# Step 5 Contents

| Strand/topic | Page | Hours | Due date | | | | Feedback |
|---|---|---|---|---|---|---|---|
| **1 Number** | | | | | | | |
| 1.1 Multiplying and dividing decimals | 6 | 1 | | | | | |
| 1.2 Multiplying and dividing fractions | 10 | 1 | | | | | |
| 1.3 Rounding and estimates | 14 | 1 | | | | | |
| 1.4 Errors | 16 | 1 | | | | | |
| **2 Algebra** | | | | | | | |
| 2.1 Expanding brackets | 20 | 1 | | | | | |
| 2.2 Sequences | 22 | 1 | | | | | |
| 2.3 Quadratic graphs | 26 | 1 | | | | | |
| 2.4 Solving simultaneous equations | 32 | 1 | | | | | |
| 2.5 Formulae | 38 | 1 | | | | | |
| **3 Ratio, proportion and rates of change** | | | | | | | |
| 3.1 Inverse proportion | 41 | 1 | | | | | |
| 3.2 Proportional change | 43 | 1 | | | | | |
| 3.3 Compound units | 45 | 1 | | | | | |

| Strand/topic | Page | Hours | Due date | | | | Feedback |
|---|---|---|---|---|---|---|---|
| **4 Geometry and measures** | | | | | | | |
| 4.1   Pythagoras' theorem | 48 | 1 | | | | | |
| 4.2   Area | 52 | 1 | | | | | |
| 4.3   Prisms and cylinders | 55 | 1 | | | | | |
| **5 Statistics** | | | | | | | |
| 5.1   Scatter graphs | 59 | 1 | | | | | |
| 5.2   Grouped data | 63 | 1 | | | | | |
| 5.3   Comparing distributions | 66 | 1 | | | | | |
| Mental maths warm-ups | | | | | | | |
| Record of achievement certificate | | | | | | | |

# Number

## 1.1 Multiplying and dividing decimals

### I can

- use place value to help multiply and divide decimals

> **Example**
>
> Work these out, without using a calculator.
>
> **a** $75 \times 0.3$    **b** $1.3 \times 0.6$
>
> **Solution**
>
> **a** If you ignore the decimal point, you have:
>
> $75 \times 3 = 225$
>
> $3 \div 10 = 0.3$
>
> You are multiplying two numbers together. If you divide one of them by 10, you divide the answer by 10.
>
> $75 \times 0.3 = 225 \div 10$
>
> $\qquad\quad = 22.5$
>
> **b** Ignore the decimal points.
>
> $13 \times 6 = 78$
>
> In the question, both numbers have been divided by 10.
>
> So divide the answer by $10 \times 10 = 100$.
>
> $1.3 \times 0.6 = 78 \div 100$
>
> $\qquad\quad = 0.78$

## Example

Work these out, without using a calculator.

**a** 75 ÷ 0.3    **b** 3 ÷ 0.04

**Solution**

**a** Writing the question without the decimal point you have:

75 ÷ 3 = 25

If you divide the divisor by 10, you multiply the answer by 10.

75 ÷ 0.3 = 25 × 10

= 250

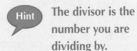

**Hint** The divisor is the number you are dividing by.

**b** Writing the question without the decimal point you have:

3 ÷ 4 = 0.75

If you divide the divisor by 100, you multiply the answer by 100.

3 ÷ 0.04 = 0.75 × 100

= 75

## Practice questions

Do not use a calculator for these questions.

**1** Work these out.

**a** 12 × 0.6 = _____

**b** 12 × 0.06 = _____

**c** 1.2 × 6 = _____

**d** 1.2 × 0.6 = _____

**2** Work these out.

**a** 3.5 × 4 = _____

**b** 3.5 × 0.4 = _____

**c** 0.4 × 0.35 = _____

**d** 0.04 × 35 = _____

**3** Work these out.

**a** $0.8 \times 0.5 =$ _____

**b** $11 \times 0.03 =$ _____

**c** $2.5 \times 0.6 =$ _____

**d** $4.1 \times 0.3 =$ _____

**4** Work these out.

**a** $0.7 \times 0.7 =$ _____

**b** $30 \times 0.3 =$ _____

**c** $5.2 \times 0.5 =$ _____

**d** $5.5 \times 0.2 =$ _____

**5** $48 \times 85 = 4080$

Use this fact to work these out.

**a** $48 \times 8.5 =$ _____

**b** $4.8 \times 8.5 =$ _____

**c** $0.48 \times 85 =$ _____

**d** $48 \times 0.85 =$ _____

**6** Work these out.

**a** $24 \div 4 =$ _____

**b** $24 \div 0.4 =$ _____

**c** $24 \div 0.04 =$ _____

**d** $2.4 \div 4 =$ _____

**7** Work these out.

**a** $36 \div 0.6 =$ _____

**b** $4.2 \div 6 =$ _____

**c** $5.4 \div 0.6 =$ _____

**d** $30 \div 0.06 =$ _____

**8** Work these out.

**a** $20 \div 0.4 =$ _____

**b** $40 \div 0.2 =$ _____

**c** $7 \div 0.5 =$ _____

**d** $7 \div 0.2 =$ _____

**9** Work these out.

**a** $150 \div 0.5 =$ _____

**b** $80 \div 0.2 =$ _____

**c** $6 \div 0.4 =$ _____

**d** $10 \div 0.04 =$ _____

**10** $648 \div 24 = 27$

Use this fact to work these out.

**a** $648 \div 2.4 =$ _____

**b** $648 \div 0.24 =$ _____

**c** $64.8 \div 2.4 =$ _____

**d** $6.48 \div 2.4 =$ _____

**11** Work these out.

**a** $12 \times 0.8 =$ _____

**b** $12 \div 0.8 =$ _____

**c** $1.2 \times 0.8 =$ _____

**d** $1.2 \div 0.8 =$ _____

**12** Work these out.

**a** $44 \times 0.4 =$ _____

**b** $44 \div 0.4 =$ _____

**c** $4.4 \times 0.4 =$ _____

**d** $4.4 \div 0.4 =$ _____

# 1.2 Multiplying and dividing fractions

## I can

* multiply and divide fractions or mixed numbers

## Example

Work these out.

**a** $\frac{2}{3} \times \frac{4}{5}$     **b** $\frac{2}{3} \div \frac{4}{5}$     **c** $1\frac{2}{3} \times 1\frac{1}{2}$     **d** $1\frac{2}{3} \div 1\frac{1}{2}$

**Solution**

**a** To multiply two fractions, multiply the numerators and multiply the denominators.

$$\frac{2}{3} \times \frac{4}{5} = \frac{2 \times 4}{3 \times 5}$$
$$= \frac{8}{15}$$

**b** To divide by a fraction, multiply by the inverse. The inverse of $\frac{4}{5}$ is $\frac{5}{4}$.

$$\frac{2}{3} \div \frac{4}{5} = \frac{2}{3} \times \frac{5}{4}$$
$$= \frac{2 \times 5}{3 \times 4}$$
$$= \frac{10}{12} \qquad \text{Simplify by dividing each number by 2.}$$
$$= \frac{5}{6}$$

Look at the expression $\frac{2 \times 5}{3 \times 4}$ above.

The 2 and the 4 have a common factor of 2 so you can divide them both by this before you multiply.

$$\frac{{}^1\cancel{2} \times 5}{3 \times \cancel{4}_2} = \frac{5}{6}$$

In this case you get the answer directly. It cannot be simplified, like $\frac{10}{12}$ can.

**c** $1\frac{2}{3} \times 1\frac{1}{2} = \frac{5}{\cancel{3}_1} \times \frac{\cancel{3}}{2} = \frac{5}{2} = 2\frac{1}{2}$

Notice that the two 3s have been divided by 3.

If you do not do this you will get the answer $\frac{15}{6}$ for the multiplication.

You then need to simplify this to $2\frac{1}{2}$.

**d** Change the divisor to an improper fraction and then multiply by its inverse.

$$1\frac{2}{3} \div 1\frac{1}{2} = \frac{5}{3} \div \frac{3}{2}$$
$$= \frac{5}{3} \times \frac{2}{3}$$
$$= \frac{5 \times 2}{3 \times 3}$$
$$= \frac{10}{9}$$
$$= 1\frac{1}{9}$$

# Practice questions

 **1** Work these out.

**a** $\frac{1}{3} \times \frac{2}{3} =$ _____

**b** $\frac{2}{5} \times \frac{2}{3} =$ _____

**c** $\frac{3}{5} \times \frac{1}{2} =$ _____

**d** $\frac{3}{4} \times \frac{3}{4} =$ _____

**2** Work these out. Write each answer as simply as possible.

**a** $\frac{1}{3} \times \frac{3}{4} =$ _____

**b** $\frac{2}{5} \times \frac{1}{4} =$ _____

**c** $\frac{2}{3} \times \frac{5}{8} =$ _____

**d** $\frac{5}{6} \times \frac{3}{4} =$ _____

**3** Work these out. Write each answer as simply as possible.

**a** $\frac{3}{5} \times \frac{5}{6} =$ _____

**b** $\frac{2}{5} \times \frac{5}{8} =$ _____

**c** $\frac{7}{8} \times \frac{4}{5} =$ _____

**d** $\frac{9}{10} \times \frac{2}{3} =$ _____

**4** Work these out.

**a** $1\frac{1}{2} \times 1\frac{1}{2} =$ _____

**b** $1\frac{1}{2} \times 2\frac{1}{2} =$ _____

**c** $1\frac{1}{3} \times 2\frac{1}{2} =$ _____

**d** $1\frac{2}{3} \times 1\frac{2}{3} =$ _____

**5** Work these out.

**a** $\frac{1}{3} \div \frac{1}{4} =$ _____

**b** $\frac{2}{3} \div \frac{3}{4} =$ _____

**c** $\frac{3}{8} \div \frac{1}{3} =$ _____

**d** $\frac{5}{6} \div \frac{2}{3} =$ _____

**6** Work these out.

**a** $\frac{1}{5} \div \frac{2}{3} =$ _____

**b** $\frac{2}{3} \div \frac{1}{5} =$ _____

**c** $\frac{3}{8} \div \frac{3}{5} =$ _____

**d** $\frac{3}{5} \div \frac{3}{8} =$ _____

**7** Work these out.

**a** $1\frac{1}{3} \div \frac{1}{2} =$ _____

**b** $1\frac{1}{2} \div \frac{1}{3} =$ _____

**c** $\frac{3}{4} \div 1\frac{1}{2} =$ _____

**d** $\frac{3}{5} \div 2\frac{1}{2} =$ _____

**8** Work these out.

**a** $1\frac{1}{2} \div 1\frac{3}{4} =$ _____

**b** $1\frac{3}{4} \div 1\frac{1}{4} =$ _____

**c** $2\frac{1}{2} \div 1\frac{1}{2} =$ _____

**d** $3\frac{1}{3} \div 2\frac{1}{2} =$ _____

**9** Work these out. Write each answer as simply as possible.

**a** $\frac{2}{3} \times \frac{1}{2} =$ _____

**b** $\frac{2}{3} \div \frac{1}{2} =$ _____

**c** $\frac{1}{2} \times \frac{2}{3} =$ _____

**d** $\frac{1}{2} \div \frac{2}{3} =$ _____

**10** Work these out. Write each answer as simply as possible.

**a** $3\frac{1}{2} \times 1\frac{1}{2} =$ _____

**b** $3\frac{1}{2} \div 1\frac{1}{2} =$ _____

**c** $1\frac{1}{2} \times 3\frac{1}{2} =$ _____

**d** $1\frac{1}{2} \div 3\frac{1}{2} =$ _____

# 1.3 Rounding and estimates

## I can

* round numbers to make estimates of calculations

## Practice questions

Do not use a calculator for these questions.

**1** Round each number to one significant figure.

    **a** $329 \approx$ _____            **b** $27.32 \approx$ _____

    **c** $0.716 \approx$ _____        **d** $5678 \approx$ _____

**2** Round each number to one significant figure to find an estimate of the answer.

    **a** $32.5 \times 61.8 \approx 30 \times$ _____ $=$ _____     **b** $491 \times 2.15 \approx$ _____ $\times$ _____ $=$ _____

    **c** $72.9 \times 3.14 \approx$ _____ $\times$ _____ $=$ _____     **d** $0.471 \times 8.36 \approx$ _____ $\times$ _____ $=$ _____

    **e** $7230 \times 37 \approx$ _____ $\times$ _____ $=$ _____     **f** $0.81 \times 0.792 \approx$ _____ $\times$ _____ $=$ _____

**3** Round each number to one significant figure to find an estimate of the answer.

**a** 432 ÷ 38 ≈ _____ ÷ _____ = _____    **b** 7.93 ÷ 4.09 ≈ _____ ÷ _____ = _____

**c** 691 ÷ 0.97 ≈ _____ ÷ _____ = _____    **d** 6.14 ÷ 0.518 ≈ _____ ÷ _____ = _____

**e** 82.3 ÷ 97.2 ≈ _____ ÷ _____ = _____    **f** 2700 ÷ 30.6 ≈ _____ ÷ _____ = _____

**4** Estimate the answer to each calculation.

**a** $\dfrac{27.2 \times 41.2}{5.91}$ ≈ _____    **b** $\dfrac{8.2 \times 113}{22.6}$ ≈ _____

**5**

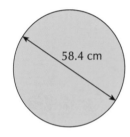

58.4 cm

**a** The circumference of a circle with a diameter of 58.4 cm is $3.14159 \times 58.4$ cm.

Estimate this length.    _____ cm

**b** The area of a circle with a diameter of 58.4 cm is $3.14159 \times 29.2 \times 29.2$ cm$^2$.

Estimate this area.    _____ cm$^2$

**6** The Body Mass Index of a man with a mass of 81.4 kg and a height of 1.91 m is $\dfrac{81.4}{1.91 \times 1.91}$.

Estimate the value of the Body Mass Index.    _____

# 1.4 Errors

## I can

- use interval notation to express possible errors

## Example

The lengths of the sides of a rectangle are 14 cm and 20 cm, to the nearest centimetre.

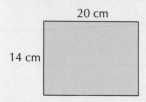

Work out an interval estimate for:

**a** the perimeter of the rectangle   **b** the area of the rectangle.

### Solution

**a** The width is between 13.5 cm and 14.5 cm so you can write:

13.5 cm $\leqslant$ width < 14.5 cm.

Notice the signs: 13.5 rounds up to 14 and 14.5 rounds up to 15.

Similarly:

19.5 cm $\leqslant$ length < 20.5 cm

The perimeter = 2(width + length) so:

2(13.5 + 19.5) $\leqslant$ perimeter < 2(14.5 + 20.5)

So 66 cm $\leqslant$ perimeter < 70 cm.

**b** The area = width × length so:

13.5 × 19.5 $\leqslant$ area < 14.5 × 20.5

So 263.25 cm$^2$ $\leqslant$ area < 297.25 cm$^2$.

## Practice questions

**1**  Write interval estimates for each of these measurements.

**a**  A length of 35 cm to the nearest centimetre  _____ ≤ length < _____

**b**  A length of 620 m to the nearest ten metres  _____ ≤ length < _____

**c**  A length of 7.4 cm to the nearest millimetre  _____ ≤ length < _____

**d**  A mass of 70 kg to the nearest kilogram  _____ ≤ mass < _____

**e**  A mass of 70 kg to the nearest ten kilograms  _____ ≤ mass < _____

**f**  A capacity of 50 ml to the nearest millilitre  _____ ≤ capacity < _____

**g**  A capacity of 50 ml to the nearest ten millilitres  _____ ≤ capacity < _____

**2**  The sides of these regular polygons are given to the accuracy shown.

Work out an interval estimate for the perimeter of each one.

**a**

7 cm to the
nearest centimetre

_____ ≤ perimeter < _____

**b**

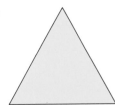

120 cm to the
nearest 10 centimetres

_____ ≤ perimeter < _____

**c**

6.2 cm to the
nearest millimetre

_____ ≤ perimeter < _____

**d**

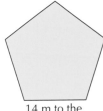

14 m to the
nearest metre

_____ ≤ perimeter < _____

**3**  **a**  The side of a square is 8 cm long, to the nearest centimetre.

Work out an interval estimate for the area of the square.

_____cm$^2$ ≤ area < _____cm$^2$

**b**  The side of a square is 8.0 cm long, to the nearest millimetre.

Work out an interval estimate for the area of the square.

_____cm$^2$ ≤ area < _____cm$^2$

**4**  This is a plan of a football pitch. The lengths of the sides are given to the nearest metre.

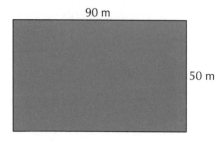

90 m

50 m

**a**  Work out an interval estimate for the perimeter of the pitch.

_____m ≤ perimeter < _____m

**b**  Work out an interval estimate for the area of the pitch.

_____m$^2$ ≤ area < _____m$^2$

# Comments, next steps, misconceptions

# 2 Algebra

## 2.1 Expanding brackets

### I can

- expand the product of two brackets

---

**Example**

Expand each expression.

**a** $(x + 3)(x - 4)$    **b** $(x - 5)^2$

**Solution**

**a** Multiply each term in the first set of brackets by each term in the second set.

$$(x + 3)(x - 4) = x^2 - 4x + 3x - 12$$

$$= x^2 - x - 12 \qquad \text{Combining the } x \text{ terms.}$$

You can write the multiplication in a table like this, if you wish.

|     | $x$    | $+3$   |
| --- | ------ | ------ |
| $x$  | $x^2$  | $+3x$  |
| $-4$ | $-4x$  | $-12$  |

**b** $(x - 5)^2 = (x - 5)(x - 5)$

$$(x - 5)(x - 5) = x^2 - 5x - 5x + 25 \qquad \text{Note that } -5 \times -5 = 25.$$

$$= x^2 - 10x + 25$$

---

## Practice questions

**1** Expand each expression.

**a** $(x + 3)(x + 5) =$ 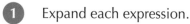 _____

**b** $(x + 4)(x + 2) =$ _____

**c** $(x + 6)(x + 1) =$ _____

**d** $(x + 5)(x + 8) =$ _____

**2** Expand each expression.

**a** $(x + 3)(x - 5) =$ _____

**b** $(x + 7)(x - 2) =$ _____

**c** $(x - 4)(x + 5) =$ _____

**d** $(x - 8)(x + 6) =$ _____

**3** Expand each expression.

**a** $(x - 3)(x - 4) =$ _____

**b** $(x - 10)(x - 2) =$ _____

**c** $(x - 6)(x - 1) =$ _____

**d** $(x - 5)(x - 4) =$ _____

**4** Expand each expression.

**a** $(x + 2)^2 =$ _____

**b** $(x + 7)^2 =$ _____

**c** $(x - 4)^2 =$ _____

**d** $(x - 10)^2 =$ _____

**5** Expand each expression.

**a** $(x - 5)(x + 6) =$ _____

**b** $(x - 9)(x - 4) =$ _____

**c** $(x - 1)^2 =$ _____

**d** $(x + 7)(x + 3) =$ _____

**e** $(x + 8)(x - 12) =$ _____

**f** $(x + 12)^2 =$ _____

# 2.2 Sequences

## I can

- recognise sequences and find further terms

---

### Example

Work out the next two terms in each sequence.

**a** 3   6   12   24   48   __   __         **b** 2   8   18   32   __   __

**c** 2   5   7   12   19   31   __   __

### Solution

**a** You have already met arithmetic sequences, where the difference between terms is constant.

This is not an arithmetic sequence.

In this sequence each term is double the previous one.

The next two terms are $48 \times 2 = 96$ and $96 \times 2 = 192$.

**b** Look at the differences.

$$2 \qquad\quad 8 \qquad\quad 18 \qquad\quad 32 \qquad\quad \underline{\phantom{00}}$$
$$\quad\; 6 \qquad\quad 10 \qquad\quad 14$$

The next differences will be 18 and 22.

The next two terms are $32 + 18 = 50$ and $50 + 22 = 72$.

You might also notice that each term is double a square number:

$2 \times 1 = 2$, $2 \times 4 = 8$, and so on.

The next two will be $2 \times 25 = 50$ and $2 \times 36 = 72$ as before.

**c** This is an example of a Fibonacci sequence. Each term is the sum of the previous two.

$2 + 5 = 7$, $5 + 7 = 12$ and so on.

The next two are $19 + 31 = 50$ and $31 + 50 = 81$.

# Practice questions

**1** Work out the next term in each sequence.

   **a** 10   16   22   28   34   ____

   **b** 2   4   8   16   32   ____

   **c** 1   3   9   27   81   ____

   **d** 120   60   30   15   ____

**2** Work out the next term in each sequence.

   **a** 3   12   27   48   75   ____

   **b** 24   12   6   3   ____

   **c** 0.5   2   4.5   8   12.5   ____

   **d** 640   160   40   10   ____

**3** Work out the missing terms in each sequence.

   **a** 3   ____   11   15   19   ____   27

   **b** 5   ____   20   40   80   ____

   **c** ____   24   12   6   3   ____

   **d** ____   600   300   150   ____   37.5

**4** **a** Work out the number of dots in each of the next two patterns.

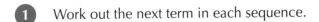

   1          3          6          10          ____     ____

   **b** Work out the next two numbers in each sequence. (Hint: Use the sequence in part **a**.)

   **i** 3   5   8   12   ____   ____          **ii** 3   9   18   30   ____   ____

**5** Work out the next two terms in each sequence.

    **a** 1  1  2  3  5  8  13  ____  ____

    **b** 1  5  6  11  17  28  ____  ____

**6** Work out the next two fractions in each sequence.

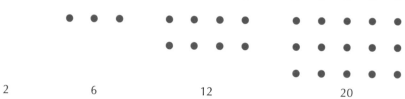

**a** $\dfrac{2}{3}$   $\dfrac{3}{5}$   $\dfrac{4}{7}$   $\dfrac{5}{9}$   $\dfrac{6}{11}$   $\dfrac{\square}{\square}$   $\dfrac{\square}{\square}$

**b** $\dfrac{1}{2}$   $\dfrac{3}{4}$   $\dfrac{5}{8}$   $\dfrac{7}{16}$   $\dfrac{9}{32}$   $\dfrac{\square}{\square}$   $\dfrac{\square}{\square}$

**7** Work out the number of dots in each of the next two patterns.

**a**

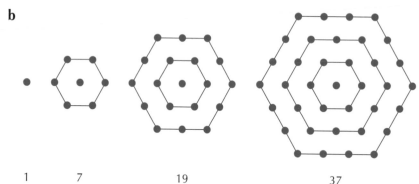

    2        6        12        20        ____  ____

**b**

    1    7        19        37        ____  ____

**8** Here is a sequence of growing patterns.

1    2    3    4

  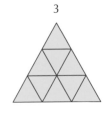 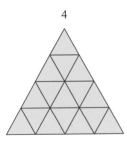

There is one small triangle in pattern 1 and four small triangles in pattern 2.

**a** How many small triangles are there in patterns 3 and 4?    _____ and _____

**b** Work out the number of small triangles in pattern 10.    _____

**9** Here is a sequence of rectangles.

Rectangle 1        Rectangle 2        Rectangle 3        Rectangle 4

 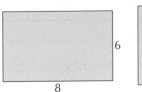 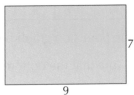

Work out the area of the next rectangle in the pattern.    _____

# 2.3 Quadratic graphs

## I can

- draw a quadratic graph
- use a quadratic graph to solve an equation

### Example

**a** Draw the graph of $y = x^2 - 3x$ for $-2 \leqslant x \leqslant 5$.

**b** Use the graph to find the approximate solutions of the equation $x^2 - 3x = 6$.

### Solution

**a** Start with a table of values. It is easier to take positive values of $x$ first.

Look for symmetry in the values in the table.

| $x$ | -2 | -1 | 0 | 1 | 2 | 3 | 4 | 5 |
|---|---|---|---|---|---|---|---|---|
| $x^2 - 3x$ | 10 | 4 | 0 | -2 | -2 | 0 | 4 | 10 |

Plot the points on graph paper. Join them with a smooth curve. Use a pencil, never a pen.

Make sure that the bottom of the graph is a curve, not a straight line.

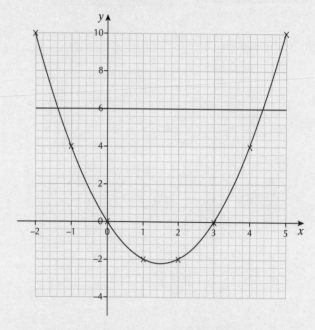

**b** Draw the straight line $y = 6$ on the graph.

Read the $x$-coordinates where the line crosses the curve.

There are two solutions, $x = 4.4$ and $x = -1.4$.

These are approximate solutions because they are read from a graph.

# Practice questions

**1**  **a** Complete this table of values for $x^2 - 4$.

| $x$ | −3 | −2 | −1 | 0 | 1 | 2 | 3 |
|---|---|---|---|---|---|---|---|
| $x^2 - 4$ | | | −3 | | | | 5 |

**b** Draw a graph of $y = x^2 - 4$ for $-3 \leqslant x \leqslant 3$.

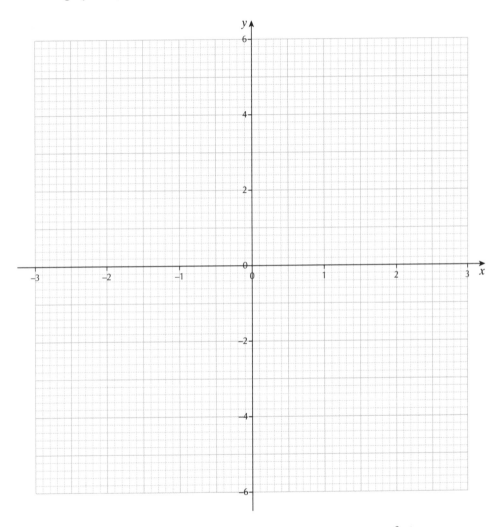

**c** Use your graph to find approximate solutions of the equation $x^2 - 4 = 2$.

$x =$ _____

**2**  **a** Complete this table of values for $6 - x^2$.

| $x$ | $-3$ | $-2$ | $-1$ | $0$ | $1$ | $2$ | $3$ |
|---|---|---|---|---|---|---|---|
| $6 - x^2$ | | | | $6$ | | | $-3$ |

**b** Draw a graph of $y = 6 - x^2$ for $-3 \leqslant x \leqslant 3$.

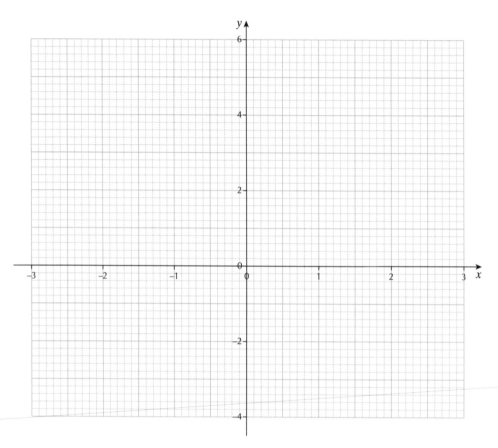

**c** Use your graph to find approximate solutions for the equation $6 - x^2 = 4$.

$x = $ _____

**d** Use your graph to find approximate solutions for the equation $6 - x^2 = -2$.

$x = $ _____

**3**    **a** Complete this table of values for $(x - 2)(x + 1)$.

| $x$ | $-3$ | $-2$ | $-1$ | $0$ | $1$ | $2$ | $3$ | $4$ |
|---|---|---|---|---|---|---|---|---|
| $(x - 2)(x + 1)$ | | 4 | | | $-2$ | | | |

**b** Draw a graph of $y = (x - 2)(x + 1)$ for $-3 \leqslant x \leqslant 4$.

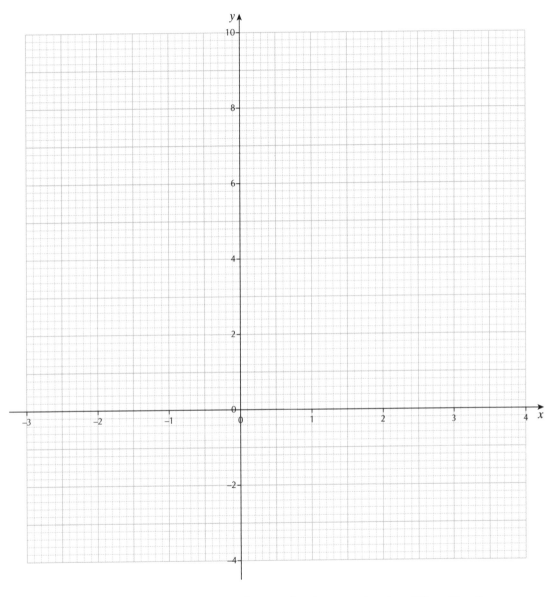

**c** Use your graph to find approximate solutions for the equation $(x - 2)(x + 1) = 6$.

$x = $ _____

**4**  **a** Complete this table of values for $x^2 + x + 2$.

| $x$ | $-3$ | $-2$ | $-1$ | $0$ | $1$ | $2$ |
|---|---|---|---|---|---|---|
| $x^2 + x + 2$ | | 4 | | | | |

**b** Draw a graph of $y = x^2 + x + 2$ for $-3 \leqslant x \leqslant 2$.

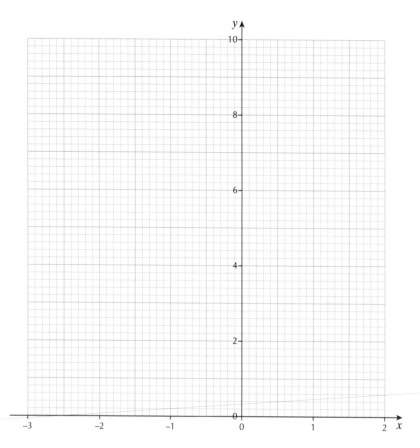

**c** Use your graph to find approximate solutions for the equation $x^2 + x + 2 = 5$.

$x = $ _____

**5**  **a** Complete this table of values for $x^2 - x - 5$.

| $x$ | $-3$ | $-2$ | $-1$ | $0$ | $1$ | $2$ | $3$ | $4$ |
|---|---|---|---|---|---|---|---|---|
| $x^2 - x - 5$ | 7 | | $-3$ | | | | | |

**b** Draw a graph of $y = x^2 - x - 5$ for $-3 \leqslant x \leqslant 4$.

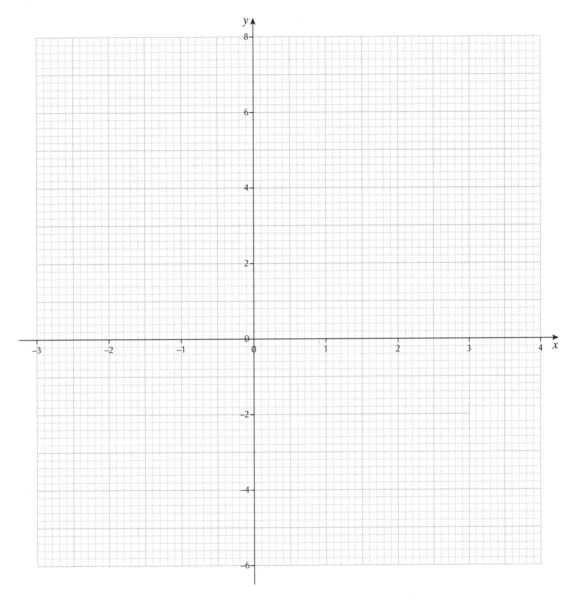

**c** Use your graph to find approximate solutions for the equation $x^2 - x - 5 = -4$.

$x =$ _____

# 2.4 Solving simultaneous equations

## I can

- use a graph to solve simultaneous linear equations

**a** Draw a graph of $2x + y = 7$.

**b** On the same axes draw a graph of $y = 2x + 1$.

**c** Use your graphs to solve simultaneously the equations $2x + y = 7$ and $y = 2x + 1$.

**Solution**

**a** It is useful to have a table of values.

For an equation such as $2x + y = 7$ choose $x = 0$ for one value and $y = 0$ for another.

| $x$ | 0 | 3.5 | 3 |
|---|---|---|---|
| $y$ | 7 | 0 | 1 |

Two pairs of values are enough to draw a straight line but it is wise to check with a third pair.

**b** For an equation such as $y = 2x + 1$ choose $x$ values and work out the $y$ values.

| $x$ | 0 | 2 | 3 |
|---|---|---|---|
| $y$ | 1 | 5 | 7 |

Finding three values is sufficient but you can do more if you wish.

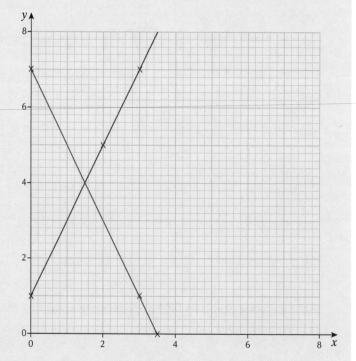

**c** The simultaneous solution is the point where the lines cross.

The coordinates of this point are $x = 1.5$ and $y = 4$.

# Practice questions

  Use these graphs to solve these simultaneous equations.

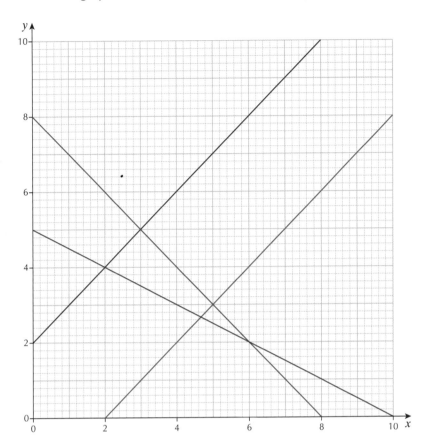

**a** $x + 2y = 10$ and $y = x + 2$ $\qquad$ $x = $ _____ and $y = $ _____

**b** $x + y = 8$ and $y = x - 2$ $\qquad$ $x = $ _____ and $y = $ _____

**c** $x + 2y = 10$ and $x + y = 8$ $\qquad$ $x = $ _____ and $y = $ _____

**2**

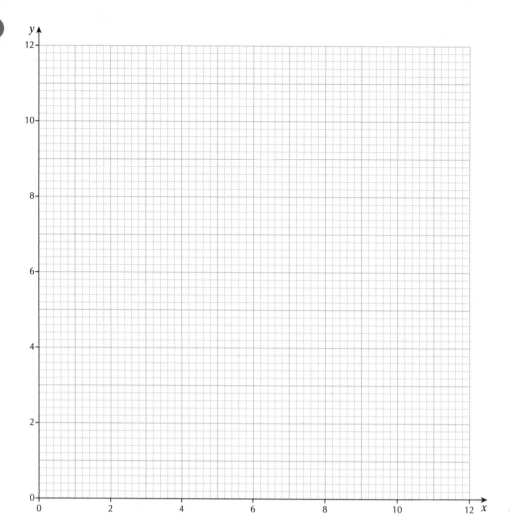

**a** Draw a graph of $x + 2y = 12$.

**b** Draw a graph of $y = \frac{1}{2}x + 1$.

**c** Use your graphs to solve simultaneously $x + 2y = 12$ and $y = \frac{1}{2}x + 1$.

$$x = \rule{2cm}{0.4pt} \text{ and } y = \rule{2cm}{0.4pt}$$

**3**

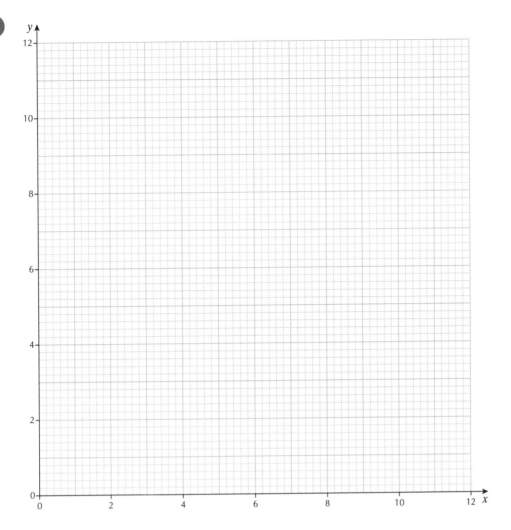

**a** Draw a graph of $3x + 2y = 24$.

**b** Draw a graph of $y = 1.5x + 3$.

**c** Use your graphs to solve simultaneously $3x + 2y = 24$ and $y = 1.5x + 3$.

$$x = \text{\underline{\hspace{2cm}}} \text{ and } y = \text{\underline{\hspace{2cm}}}$$

**4**

**a** Draw a graph of $x + y = 25$.

**b** Draw a graph of $y = x - 5$.

**c** Use your graphs to solve simultaneously
$x + y = 25$ and $y = x - 5$.

$$x = \text{_____} \text{ and } y = \text{_____}$$

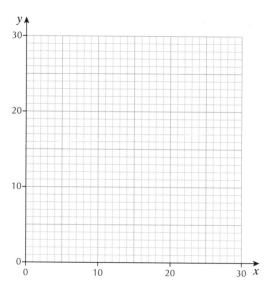

**5**    Use these graphs to solve the simultaneous equations.

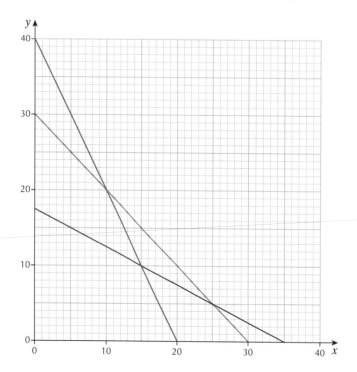

**a**  $x + y = 30$ and $2x + y = 40$          $x = \text{_____}$ and $y = \text{_____}$

**b**  $x + 2y = 35$ and $2x + y = 40$          $x = \text{_____}$ and $y = \text{_____}$

**6** Here are two equations.

$3x + 5y = 150$ $y = x - 10$

Draw graphs to solve the equations simultaneously.

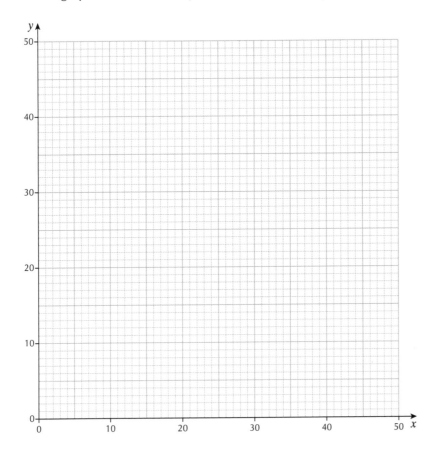

$x =$ _____ and $y =$ _____

# 2.5 Formulae

## I can

- substitute numbers into expressions and formulae

A formula for the area, $A$, of this trapezium is:

$A = \frac{1}{2}(a + b)h$

Work out the value of $A$ if $a = 10$, $b = 12$ and $h = 9$.

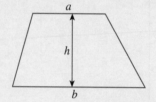

### Solution

Substitute the values into the formula. Work out the value of the term inside the brackets first.

$A = \frac{1}{2} \times (10 + 12) \times 9$

$= \frac{1}{2} \times 22 \times 9$

$= 99$

## Practice questions

**1**  Work out the value of each expression, given that $p = 5$ and $q = 7$.

    **a**  $2p + 3q = $ _____
    **b**  $4(q - p) = $ _____
    **c**  $p^2 + q^2 = $ _____

**2**  Work out the value of each expression, given that $t = 5$ and $u = 7$.

    **a**  $4 + 6t = $ _____
    **b**  $4u + 6t = $ _____
    **c**  $2u^2 + 3 = $ _____

    **d**  $5(t + u) = $ _____
    **e**  $t(u + 1) = $ _____
    **f**  $t(u - 4) = $ _____

**3**  Fill in the missing numbers in this table.

| $x$ | 3 | 5 | 7 | 1 | −4 |
|-----|----|----|----|----|----|
| $2x^2$ | 18 | | | | |

**4** Work out the value of each expression, given that $y = 3$.

a $4y^2 =$ _____

b $(4y)^2 =$ _____

c $(4 + y)^2 =$ _____

**5** A formula used in science is $E = IR$.

a Work out the value of $E$ if $I = 2$ and $R = 30$.

$E =$ _____

b Work out the value of $E$ if $I = 0.1$ and $R = 12$.

$E =$ _____

**6** Work out the value of each expression, given that $t = 9$.

a $\sqrt{t} =$ _____

b $\sqrt{4t} =$ _____

c $4\sqrt{t} =$ _____

d $\sqrt{t} + 7 =$ _____

e $\sqrt{t + 7} =$ _____

f $3\sqrt{t} - 5 =$ _____

**7** A formula used in science is $v = u + at$.

Work out the value of $v$ when $u = 20$, $a = 0.5$ and $t = 12$.

$v = u + at =$ _____

**8** Here is a temperature formula.

$F = 1.8C + 32$

a Work out the value of $F$ when $C = 10$.

$F = 1.8C + 32 =$ _____

b Work out the value of $F$ when $C = -10$.

$F = 1.8C + 32 =$ _____

**9** A formula used in science is $s = \left(\frac{u + v}{2}\right)t$.

Work out the value of $s$ when $u = 5$, $v = 25$ and $t = 6$.

$s = \left(\frac{u + v}{2}\right)t =$ _____

**10** A formula for the total surface area, $A$, of this cube is $A = 2(ab + ac + bc)$.

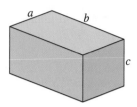

Work out the value of $A$ when $a = 4$, $b = 5$ and $c = 3$.

$A = 2(ab + ac + bc) = $ _____

---

## Comments, next steps, misconceptions

⬤ ⬤ ⬤

☐ ☐ ☐

☐ ☐ ☐

☐ ☐ ☐

☐ ☐ ☐

☐ ☐ ☐

☐ ☐ ☐

☐ ☐ ☐

☐ ☐ ☐

☐ ☐ ☐

☐ ☐ ☐

☐ ☐ ☐

☐ ☐ ☐

☐ ☐ ☐

☐ ☐ ☐

# 3 Ratio, proportion and rates of change

## 3.1 Inverse proportion

### I can

- use the concept of inverse proportion to calculate missing values

---

**Example**

The time taken for a journey is inversely proportional to the speed.

At 10 mph a journey takes 4.5 hours.

How long does the same journey take at 60 mph?

**Solution**

A table of values is useful.

$\times 6$

| Speed (mph) | 10 | 60 |
|---|---|---|
| Time (h) | 4.5 | |

$\div 6$

$60 \div 10 = 6$

The speed is multiplied by 6 so the time is divided by 6.

Time $= 4.5 \div 6 = 0.9$ hours.

Notice that speed $\times$ time $= 45$ in each case.

---

## Practice questions

**1**   If Tebor walks, at 4 mph, a journey takes him 100 minutes. How long does the same journey take if Tebor cycles, at 16 mph?

_____ minutes

2  If Aimee walks, at 3 mph, a journey takes her 150 minutes. How long does it take her if she:

    **a** jogs, at 6 mph _____ minutes    **b** cycles, at 15 mph _____ minutes

    **c** drives, at 30 mph _____ minutes    **d** drives, at 60 mph? _____ minutes

3  When Lucy drives at 30 mph, it takes her 40 minutes to travel work.

    **a** How long does it take at 15 mph?           _____ minutes

    **b** How long does it take at 60 mph?           _____ minutes

    **c** How long does it take at 10 mph?           _____ minutes

4  The length of time a job takes is inversely proportional to the number of people working on it.

It takes 24 days for 10 people to complete a job. How long does the same job take for:

**a** 5 people _____ days

**b** 20 people _____ days

**c** 30 people? _____ days

5  Jasmine draws some rectangles, all with the same area. The lengths and widths are inversely proportional.

Complete this table of values.

| Length (cm) | 12.6 | 25.2 |     |     |
| ----------- | ---- | ---- | --- | --- |
| Width (cm)  | 2.4  |      | 4.8 | 0.8 |

6  Gary draws some pyramids, all with the same volume. The base areas and heights are inversely proportional.

Complete this table of values.

| Base area (cm$^2$) | 25 | 50 | 100 |    |    |
| ------------------ | -- | -- | --- | -- | -- |
| Height (cm)        | 6  |    |     | 12 | 18 |

# 3.2 Proportional change

## I can

* use proportionality to calculate missing values

---

### Example

The cost of 23 litres of petrol is £29.90. Work out the cost of 37 litres of petrol.

**Solution**

| Quantity (litres) | 37 | 23 |
|---|---|---|
| Cost (£) | £48.84 | |

The cost is proportional to the number of litres.

The multiplier for the quantity is $23 \div 37 = 0.621\ldots$

Multiply the cost by this.

$$0.621\ldots \times £48.84 = £30.36$$

Use a calculator. Keep the answer to the first calculation in your calculator. Do not round it. Use it for the second calculation.

---

## Practice questions

**1**  17 litres of petrol cost £21.93. Work out the cost of 38 litres of petrol.

£_____

**2**  350 g of carrots cost £0.63. Work out the cost of 450 g of carrots.

£_____

**3**  240 g of grapes cost £1.08. Work out the cost of:

**a**  300 g of grapes          £_____

**b**  520 g of grapes          £_____

**c**  135 g of grapes.         £_____

**4** Hassan spends £11.50 when he buys 8.91 litres of petrol. Work out how many litres he can buy for:

**a** £20 _____ litres

**b** £49.25. _____ litres

**5** In a supermarket, 0.2 g of saffron costs £1.76.

**a** Work out the cost of 0.65 g of saffron.

£ _____

**b** What mass can you buy for £20?

_____ g

**6** The price of 3.4 g of gold is £90.44.

Complete this table of prices.

| Mass (g) | 3.4 g | 8.5 g | 1.2 g | |
|---|---|---|---|---|
| Price (£) | £90.44 | | | £1000 |

**7** Before decimalisation, lengths were measured in an old unit called inches.

A length of 20 cm is approximately the same as 8 inches. Use this fact to change:

**a** 65 cm to inches _____ inches

**b** 450 cm to inches _____ inches

**c** 60 inches to centimetres. _____ cm

**8** Liquids can be measured in litres or pints.

A capacity of 6 litres is approximately equal to 10.5 pints.

**a** What capacity, in pints, is equal to 25 litres? _____ pints

**b** What capacity, in litres, is equal to 25 pints? _____ litres

# 3.3 Compound units

## I can

• understand and use compound units

---

### Example

A piece of copper has a mass of 38.8 g and a volume of 4.3 cm$^3$.

**a** Work out the density of copper.

**b** Work out the mass of 10 cm$^3$ of copper.

**Solution**

**a** Density $= \dfrac{\text{mass}}{\text{volume}}$

$\quad\;\; = \dfrac{38.8}{4.3}$        Substitute the numbers you know.

$\quad\;\; = 9.0$ g/cm$^3$     Round the answer.

The units are g/cm$^3$ because you are dividing grams by cubic centimetres.

**b** Density $= \dfrac{\text{mass}}{\text{volume}}$

$\quad 9.0 = \dfrac{\text{mass}}{10}$       Substitute the numbers you know.

$\rightarrow$ mass $= 9.0 \times 10 = 90$ g     You must give the units.

---

## Practice questions

**1** A car travels 137 km in 1 hour 45 minutes.

Work out the average speed, in km/hour.        _____ km/hour

**2** A train travels at an average speed of 96 km/hour.

    **a** How far does it travel in 45 minutes?        _____ km

    **b** How long does it take to travel 224 km?        _____

**3** Jess runs 5 kilometres in 18 minutes.

    **a** How long does she take to run 1 km?        _____ minutes

    **b** Work out her speed in metres/minute.       _____ m/min

**4** Raj jogs for 35 minutes at an average speed of 3 m/s.

    How many kilometres does he travel?       _____ km

**5** The price of a 250 g box of cereal is £1.85. Work out the cost:

    **a** per 100 g       _____

    **b** per kilogram.       _____

**6** 150 g of mixed nuts cost £2.15. Work out the cost:

    **a** per 100 g       _____

    **b** in pence/gram       _____ p/g

    **c** in gram/£.       _____ g/£

**7** 450 g of strawberries cost £1.89. Work out:

    **a** the cost per 100 g       _____

    **b** the price in gram/£.       _____ g/£

**8** A label says that the price of shower gel is 54p per 100 ml.

    Work out the cost of a 750 ml bottle.       £ _____

**9** In an experiment, 34.5 g of a substance has a volume of 13.2 cm$^3$.

    Work out the density of the substance.       _____ g/cm$^3$

**10** The density of silver is 10.5 g/cm$^3$.

    **a** Work out the mass of 4.2 cm$^3$ of silver.      _____ g

    **b** Work out the volume of 25 g of silver.      _____ cm$^3$

## Comments, next steps, misconceptions

# **4** Geometry and measures

## 4.1 Pythagoras' theorem

### I can

* use Pythagoras' theorem to work out the missing length of a side of a right-angled triangle

**Example**

Calculate the length of the third side of this triangle.

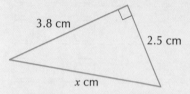

**Solution**

The square of the hypotenuse is the sum of the squares of the other two sides.

$x^2 = 2.5^2 + 3.8^2 = 20.69$       Use a calculator.

$x = \sqrt{20.69}$

    $= 4.548....$

    $= 4.5$ to one decimal place

One decimal place is sensible, because the other lengths are given to one decimal place.

# Practice questions

**1** Work out the length of the lettered side in each triangle.

**a**

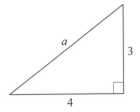

$a =$ _____

**b**

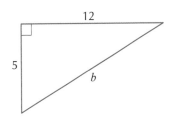

$b =$ _____

**c**

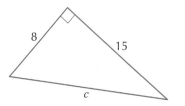

$c =$ _____

**d**

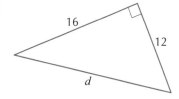

$d =$ _____

**e**

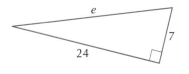

$e =$ _____

**f**

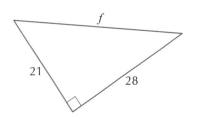

$f =$ _____

**2** The lengths of the two shorter sides of a right-angled triangle are 1.6 m and 6.3 m.

Work out the length of the longest side. _____ m

**3** The lengths of the two shorter sides of a right-angled triangle are 4.5 cm and 10.8 cm.

Work out the length of the longest side. _____ cm

**4** Work out the length of the lettered side in each triangle. Round your answers to one decimal place.

**a**

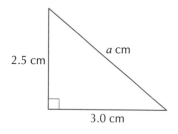

2.5 cm

a cm

3.0 cm

a = _____

**b**

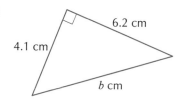

6.2 cm

4.1 cm

b cm

b = _____

**c**

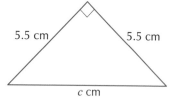

5.5 cm    5.5 cm

c cm

c = _____

**d**

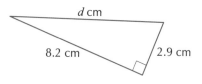

d cm

8.2 cm    2.9 cm

d = _____

**5** The length of the longest side of a right-angled triangle is 10 cm. The length of the shortest side is 6 cm.

Work out the length of the third side.

_____ cm

**6** Work out the length of the third side of each triangle.

**a**

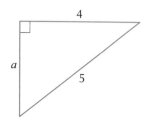

4

a

5

a = _____

**b**

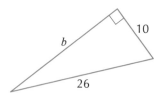

10

b

26

b = _____

**7**  Work out the length of the lettered side in each triangle. Round your answers to one decimal place.

**a**

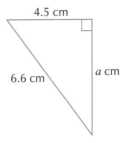

$a =$ _____

**b**

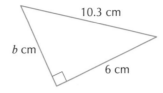

$b =$ _____

**c**

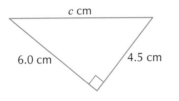

$c =$ _____

**d**

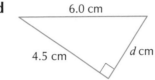

$d =$ _____

# 4.2 Area

## I can

- use the formula for the area of a trapezium
- work out the area of a compound shape by dividing it into separate parts

---

### Example

Work out the area of this shape. All the lengths are in centimetres.

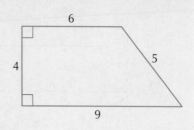

**Solution**

Here are two possible methods.

**Method 1**

The shape is a trapezium. Use the formula for the area of a trapezium.

$$\text{Area} = \frac{(6 + 9)}{2} \times 4$$

$$= 7.5 \times 4$$

$$= 30 \text{ cm}^2$$

**Method 2**

Divide the shape into a rectangle and a triangle.

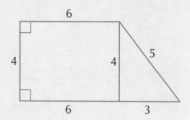

Area of rectangle = $6 \times 4 = 24$

Area of triangle = $\frac{1}{2} \times 3 \times 4 = 6$

Total area = $24 + 6 = 30$ cm$^2$   This is the same answer as before.

Notice that neither method uses the length marked 5.

# Practice questions

**1** Work out the area of each trapezium.

**a**

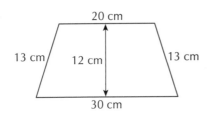

Area = _____ cm$^2$

**b**

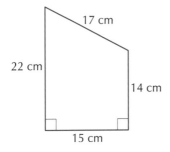

Area = _____ cm$^2$

**c**

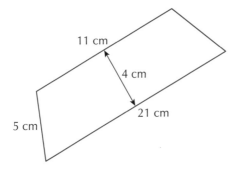

Area = _____ cm$^2$

**2** Work out the area of each shape. All lengths are in centimetres.

**a**

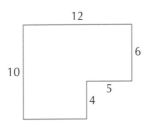

Area = _____ cm$^2$

**b**

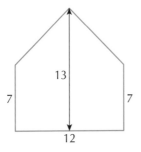

Area = _____ cm$^2$

**c**

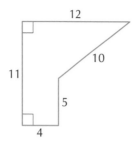

Area = _____ cm$^2$

---

**3** This shape is made from a semicircle and a rectangle.

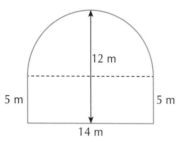

Work out its area.  Area = _____

---

**4** Work out the area of this shape.

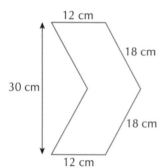

Area = _____

---

**5** This diagram shows a coloured octagon inside a rectangle.

Work out the area of the octagon. All lengths are in metres.

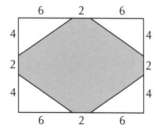

Area = _____

# 4.3 Prisms and cylinders

## I can

- work out the volume of a prism
- work out the volume of a cylinder

Work out the volume of each shape.

**a** A prism

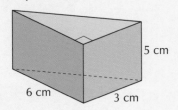

5 cm

6 cm  3 cm

**b** A cylinder

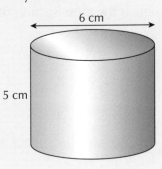

6 cm

5 cm

**Solution**

**a** The cross-section of the prism is a right-angled triangle.

Area of triangle $= \frac{1}{2} \times 6 \times 3$

$= 9 \text{ cm}^2$

3 cm

6 cm

Volume of prism = area of cross-section × length

$= 9 \times 5$

$= 45 \text{ cm}^3$

**b** The cross-section of the cylinder is a circle of diameter 6 cm.

Area of circle $= \pi \times \text{radius}^2$

$= \pi \times 3^2$

$= 9\pi \text{ cm}^2$

Volume of cylinder = area of circle × length

$= 9\pi \times 5$

$= 45\pi$

$= 141 \text{ cm}^3$ to the nearest whole number.

It is easier to leave multiplying by $\pi$ until the end.

# Practice questions

**1** Work out the volume of this prism.

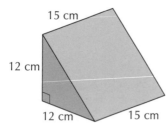

Volume = _____ cm$^3$

**2** This equilateral triangle is the cross-section of a prism.

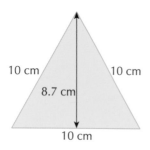

The prism is 6 cm long.

Work out the volume of the prism.

Volume = _____ cm$^3$

**3** Work out the volume of this prism. All lengths are in centimetres.

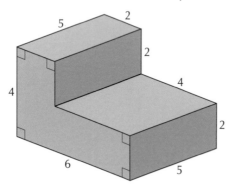

Volume = _____ cm$^3$

**4**    This cylinder is 50 cm long and has a diameter of 10 cm.

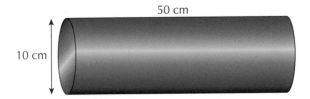

Work out the volume of the cylinder.

Volume = _____ cm³

**5**    Calculate the volume of this cylinder.

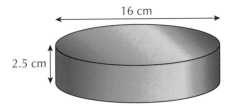

Volume = _____ cm³

**6**    This food can is 10.4 cm high.

The top is a circle of radius 3.6 cm.

Calculate the volume of the can.

Volume = _____ cm³

**7**    This is the end view of a shed.

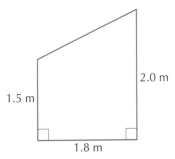

The shed is 2.4 m long.

Work out the volume of the shed.

Volume = _____ m³

# Comments, next steps, misconceptions

4 Geometry and measures

# 5 Statistics

## 5.1 Scatter graphs

### I can

- plot pairs of values on a scatter diagram
- describe the relationship between pairs of data
- draw a line of best fit

**Example**

This table shows the distance of 10 train journeys and the cost of the ticket, to the nearest pound.

| Distance (km) | 36 | 60 | 20 | 50 | 80 | 26 | 95 | 85 | 40 | 70 |
|---|---|---|---|---|---|---|---|---|---|---|
| Cost (£) | 15 | 23 | 9 | 19 | 28 | 17 | 30 | 25 | 20 | 22 |

a  Draw a scatter diagram to show the data.

b  Describe any relationship between the data.

c  Draw a line of best fit if it is sensible to do so.

**Solution**

a  Draw a pair of axes and choose sensible scales. The distance must go up to 95 and the cost to 30.

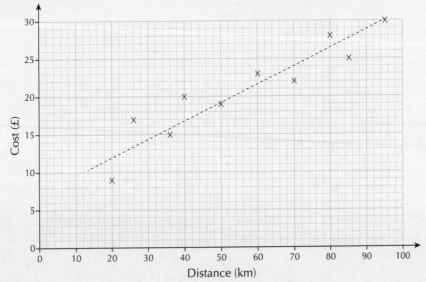

Each pair of values gives the coordinates for one point.

b  Longer distances tend to cost more than shorter ones. This is positive correlation.

c  The graph shows a line of best fit drawn by eye. It shows the trend of the points.

# Practice questions

**1** This table shows the marks of 12 people in a spelling test and in an arithmetic test.

| Spelling | 32 | 40 | 22 | 46 | 30 | 48 | 40 | 24 | 30 | 56 | 26 | 40 |
|----------|----|----|----|----|----|----|----|----|----|----|----|----|
| Arithmetic | 34 | 24 | 46 | 34 | 40 | 26 | 30 | 50 | 46 | 20 | 54 | 40 |

**a** Draw a scatter diagram to show this data.

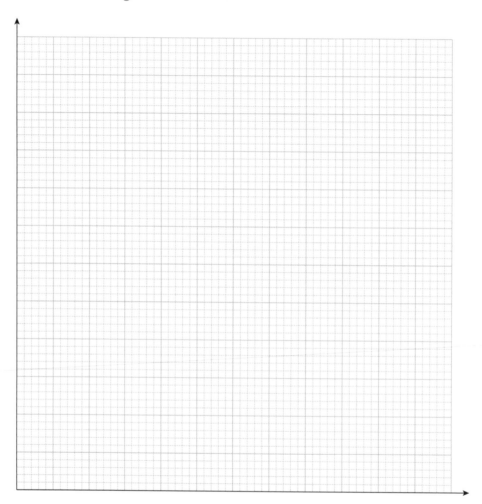

**b** Describe any relationship between the data.

_____

**c** If there is a relationship between the data, draw a line of best fit on the diagram.

**2** This table shows the ages of some adults and the time each one takes to run 100 metres.

| Age (years) | 30 | 40 | 22 | 36 | 45 | 20 | 42 | 31 | 25 | 37 | 35 | 27 |
|---|---|---|---|---|---|---|---|---|---|---|---|---|
| Time (s) | 24 | 27 | 25 | 23 | 20 | 15 | 17 | 18 | 20 | 19 | 15 | 27 |

**a** Draw a scatter diagram to show this data.

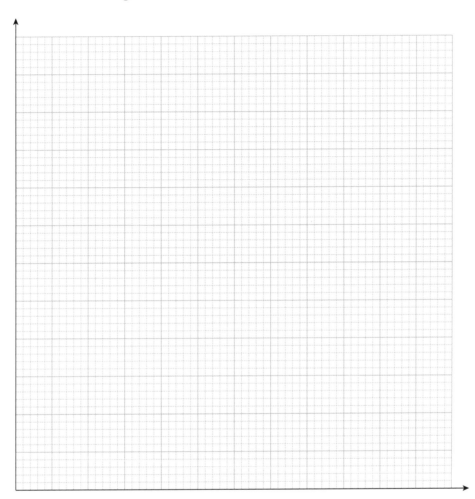

**b** Describe any relationship between the data.

_____

**c** If there is a relationship between the data, draw a line of best fit on the diagram.

**3** This table shows the midday temperature each day for ten days and the number of ice-creams sold in a shop over the same period.

| Temperature (°C) | 15 | 25 | 16 | 10 | 23 | 20 | 28 | 12 | 25 | 17 |
|---|---|---|---|---|---|---|---|---|---|---|
| Ice-creams sold | 100 | 240 | 190 | 50 | 240 | 120 | 300 | 80 | 200 | 150 |

**a** Draw a scatter diagram to show this data.

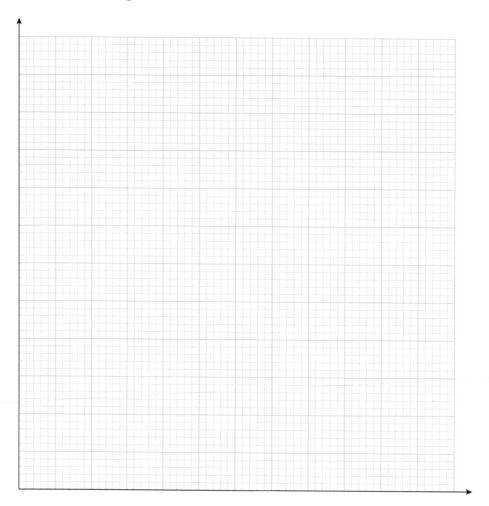

**b** Describe any relationship between the data.

_____

**c** If there is a relationship between the data, draw a line of best fit on the diagram.

# 5.2 Grouped data

## I can

- estimate the mean or the median from grouped data
- identify the modal class for grouped data

### Example

This table shows the heights of 50 plants. The heights are recorded to the nearest centimetre.

| Height (cm) | 6–10 | 11–15 | 16–20 | 21–25 | 26–30 |
|---|---|---|---|---|---|
| Frequency | 5 | 16 | 8 | 12 | 9 |

a  Identify the modal class.

b  Which class includes the median height?

c  Estimate the mean height.

**Solution**

a  The modal class is the one with the highest frequency. It is 11–15 cm.

b  There are 50 plants so the median is the mean of the 25th and 26th.

There are 5 plants in the first class, 5 + 16 = 21 in the first two classes and 21 + 8 = 29 in the first three classes. As 21 is less than 25 and 29 is more than 26, the median is in the third class: 16–20 cm.

c  Use the mid-value of each class. It is best to put the calculations in a table.

| Height | Mid-value ($m$) | Frequency ($f$) | $m \times f$ |
|---|---|---|---|
| 6–10 | 8 | 5 | 40 |
| 11–15 | 13 | 16 | 208 |
| 16–20 | 18 | 8 | 144 |
| 21–25 | 23 | 12 | 276 |
| 26–30 | 28 | 9 | 252 |
| Total | | 50 | 920 |

An estimate of the mean height is 920 ÷ 50 = 18.4 cm.

**Note:** It is an estimate because you do not know the exact height of all 50 plants.

# Practice questions

**1** This table gives the masses of 60 pebbles from a beach.

| Mass (g) | Frequency | | |
|---|---|---|---|
| 21–25 | 10 | | |
| 26–30 | 18 | | |
| 31–35 | 15 | | |
| 36–40 | 8 | | |
| 41–45 | 3 | | |
| 46–50 | 6 | | |
| Total | 60 | | |

**a** Identify the modal class. _____

**b** Which class contains the median value? _____

**c** Estimate the mean mass. You can use the empty columns in the table if you wish.

Mean = _____ g

**2** The students in a maths class see how long they can hold their breath. Here are the results.

| Time (s) | Frequency | | |
|---|---|---|---|
| 30–39 | 6 | | |
| 40–49 | 6 | | |
| 50–59 | 10 | | |
| 60–69 | 12 | | |
| 70–79 | 2 | | |
| Total | 36 | | |

**a** Which is the modal class? _____

**b** Which class contains the median value? _____

**c** Estimate the mean time for which a student can hold his or her breath. You can use the empty columns in the table if you wish.

Mean = _____ s

**3**  The table shows how many millimetres of rain there are in a town in June each year for 25 years.

| Rain (mm) | Frequency | | |
|---|---|---|---|
| At least 20 and less than 40 | 2 | | |
| At least 40 and less than 60 | 5 | | |
| At least 60 and less than 80 | 8 | | |
| At least 80 and less than 100 | 6 | | |
| At least 100 and less than 120 | 4 | | |
| Total | 25 | | |

a  Which is the modal class?  _____

b  Which class contains the median value?  _____

c  Estimate the mean amount of rainfall in June in the town. You can use the empty columns in the table if you wish.

Mean = _____ mm

**4**  A police officer records the speeds of cars on a motorway. Here are the results.

| Speed ($x$ miles per hour) | Frequency | | |
|---|---|---|---|
| $40 < x \leqslant 50$ | 3 | | |
| $50 < x \leqslant 60$ | 10 | | |
| $60 < x \leqslant 70$ | 31 | | |
| $70 < x \leqslant 80$ | 24 | | |
| $80 < x \leqslant 90$ | 9 | | |
| $90 < x \leqslant 100$ | 3 | | |

a  How many cars were in the survey?  _____

b  The speed limit on a motorway is 70 miles per hour.
   Work out the percentage of the cars that are speeding.  _____

c  Estimate the mean speed of the cars. _____ miles per hour

# 5.3 Comparing distributions

## I can

- use the mean, median, mode and range to compare distributions
- use diagrams to compare two distributions

### Example

This bar chart shows the results of a maths test and a science test.

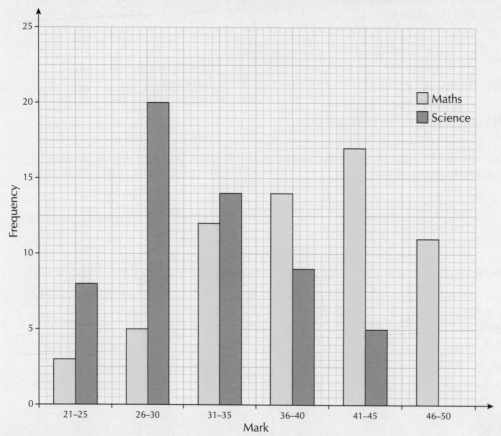

a  Which subject has the higher modal class?

b  Which subject has the higher mean?

c  Which subject has the higher range?

**Solution**

a  The modal class for maths is 41–45 and for science it is 26–30. Maths has the higher modal class.

b  Maths has the higher mean. You can tell this from the shape of the chart. Maths has more in the higher mark columns and less in the lower mark columns. You do not need to work out the mean to answer the question.

c  Maths has a higher range. The maths marks cover six classes and the science ones only five.

# Practice questions

**1**    Some plants are grown in two different soils. The bar chart shows the heights of the plants.

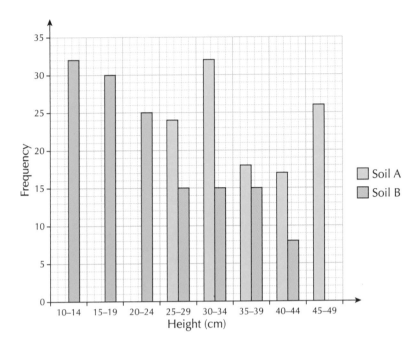

**a**  Write down the modal class for each soil.

Soil A _____ Soil B _____

**b**  Which soil has the larger mean height? Give a reason for your answer.

Soil _____          Reason _____

**c**  Which soil has the larger range? Give a reason for your answer.

Soil _____          Reason _____

**2** Here are some details about the ages of the people in three different keep-fit groups.

| Group | Members | Mean age | Median age | Youngest | Oldest |
|---|---|---|---|---|---|
| X | 25 | 32.5 | 31 | 20 | 35 |
| Y | 30 | 26.8 | 31 | 27 | 38 |
| Z | 21 | 35.9 | 31 | 26 | 37 |

**a** Which group has the largest average age? Give a reason for your answer.

Group _____        Reason _____

**b** Which group has the largest age range? Give a reason for your answer.

Group _____        Reason _____

**3**  The Panthers and the Tigers are youth football clubs.

These pie charts show the age distribution for each club.

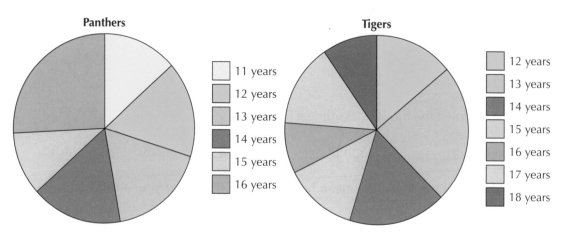

Panthers — 11 years, 12 years, 13 years, 14 years, 15 years, 16 years

Tigers — 12 years, 13 years, 14 years, 15 years, 16 years, 17 years, 18 years

**a** Write down the modal age for each club.

Panthers _____        Tigers _____

**b** Write down the median age for each club.

Panthers _____        Tigers _____

**c** Write down the age range for each club.

Panthers _____        Tigers _____

# Comments, next steps, misconceptions

## Mental warm-up 1: Number/Algebra

**1** Work out $30.5 \times 0.2$.

**2** Work out $12.4 \div 0.4$.

**3** Work out $1\frac{1}{2} \times \frac{3}{4}$.

**4** Work out $4 \div \frac{2}{3}$.

**5** Estimate the answer to $41.8 \times 29.5$.

**6** Estimate the answer to $7.81 \div 0.219$.

**7** Here is a calculator answer.

> 298328

> Round it to 2 significant figures.

**8** The measurement of a length $x$ is in the interval $9.5 \leqslant x < 10.5$.
Work out an interval for the value of $4x$.

**9** Jason finds the mass of a book, to the nearest ten grams, is 520 g. What is the smallest possible mass of the book?

**10** This is a graph of $y = x(4 - x)$.

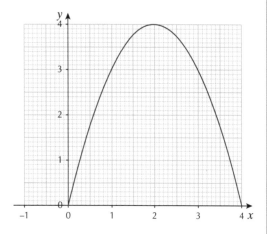

> Use the graph to solve the equation $x(4 - x) = 3$.

**11** The equation of a curve is $y = x^2 - x - 5$.
A point on the curve is $(2, \underline{\phantom{xx}})$.
Work out the missing coordinate.

**12** Expand $(x - 3)^2$.

**13** Expand $(a + 4)(a - 4)$.

## Mental warm-up 2: Algebra

**1** Here is the start of a sequence.
1   3   9   27   81   …
Write down the next term.

**2** Here are the first five terms in a sequence.
20   12   8   6   5   …
Work out the seventh term.

**3** The $n$th term of a sequence is $2n(n + 1)$.
Work out the fourth term.

**4** Here is a formula.
$a = 2b^2c$
Work out the value of $a$ when $b = 2$ and $c = 3$.

**5** $t = \sqrt{\dfrac{x}{y}}$

> Work out the value of $t$ when $x = 36$ and $y = 4$.

Use this diagram for questions 6 and 7.

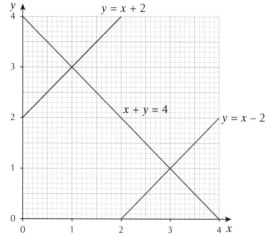

**6** Use the graph to solve the equations $x + y = 4$ and $x - y = 2$ simultaneously.

**7** Use the graph to explain why you cannot solve the equations $y = x + 2$ and $y = x - 2$ simultaneously.

**8** Work out the missing value in this table.

| $x$ | −4 | −2 | 0 | 2 | 4 |
|---|---|---|---|---|---|
| $(x - 2)(x + 3)$ | 6 | | −6 | 0 | 14 |

## Mental warm-up 3: Ratio, proportion and rates of change/ Geometry and measures

**1** The variables $x$ and $y$ are in inverse proportion.
Work out the missing value in the table.

| $x$ | 5 | 10 |
|-----|---|----|
| $y$ | 8 |    |

**2** On Monday 8000 people visit a website. On Tuesday the number increases by 50%. How many people visit the website on Tuesday?

**3** In a sale, the original price of a television is multiplied by 0.85 to work out the new price. What is the percentage reduction?

**4** A snail travels 2 metres in 8 minutes. Work out the speed of the snail in metres/minute.

**5** The cost of 250 g of apples is 90p. Work out the cost per 100g.

**6** A piece of metal has a volume of 100 cm$^3$ and a mass of 1.2 kg. Work out the density of the metal, in g/cm$^3$.

**7** Work out the length of the unlabelled side in this triangle.

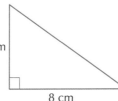
6 cm
8 cm

**8** Work out the area of this trapezium.

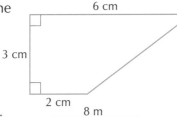
6 cm
3 cm
2 cm

**9** This shape is made from a rectangle and a triangle. Work out the area of the shape.

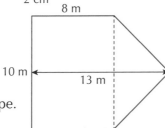
8 m
10 m
13 m

## Mental warm-up 4: Geometry and measures/Statistics

**1** This is the cross-section of a triangular prism.

5 cm
3 cm
4 cm

The prism is 20 cm long. Work out the volume of the prism.

**2** The area of the top of this cylinder is 80 cm$^2$. The volume of the cylinder is 40 cm$^3$. Work out the height of the cylinder.

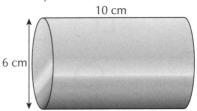

**3** Write down an expression for the volume of this cylinder in cubic centimetres (cm$^3$). Leave $\pi$ in your answer.

10 cm
6 cm

Use this table for questions 4, 5, 6 and 7.

**The lengths of 20 leaves**

| Length ($x$ cm) | Frequency ($f$) | Mid-value ($m$) | $f \times m$ |
|-----------------|-----------------|-----------------|--------------|
| $10 \leqslant x < 15$ | 4 | 12.5 | 50 |
| $15 \leqslant x < 20$ | 8 | 17.5 | 140 |
| $20 \leqslant x < 25$ | 6 | 22.5 | 135 |
| $25 \leqslant x < 30$ | 2 | 27.5 | 55 |
|  |  | Total | 380 |

**4** Work out the modal class.

**5** Estimate the mean leaf length.

**6** Explain why the range cannot be more than 20 cm.

**7** What can you say about the median leaf length?

# Record of achievement certificate

## Step 5

### Congratulations on achieving Step 5!

Name _____

Date _____

Signed _____